José MUSANGANA KAPUMBU

FIGHTING THE DEHUMANIZATION OF NATURE

José MUSANGANA KAPUMBU

FIGHTING THE DEHUMANIZATION OF NATURE

Foreword: Olivier-Jeff KALALA T.

ScienciaScripts

Imprint

Cover image: www.ingimage.com

This book is a translation from the original published under ISBN 978-620-6-72871-9.

Publisher:
Sciencia Scripts
is a trademark of
Dodo Books Indian Ocean Ltd. and OmniScriptum S.R.L publishing group

120 High Road, East Finchley, London, N2 9ED, United Kingdom
Str. Armeneasca 28/1, office 1, Chisinau MD-2012, Republic of Moldova, Europe
Managing Directors: Ieva Konstantinova, Victoria Ursu
info@omniscriptum.com

Printed at: see last page
ISBN: 978-620-8-51643-7

A :

- My dear and loving wife, Francine Kaniki,
- My beloved and adored children :

My son Marien-Jacob Lusambo Musangana and

My daughter Tshilanda Mukandayi Musangana,

I dedicate this reflection to you.

PREFACE

By virtue of its creation, extent, diversity and multifaceted nature[1] makes a major contribution to the flourishing of life, provided that its management takes into account economic, social, environmental and ethical dimensions in a balanced way.

This management of nature must be both sustainable and integrated, because it must not only ensure and improve potential but also maintain ecosystems. To achieve this, it must take account of all the functions that society expects of nature.

To ensure a balance between nature's supply and society's demand, we need to think about a strategy based on a rational analysis of natural potential and, more importantly, a method that adheres to utilitarian values and respect for nature.

To achieve this, the introspective approach, more often reserved for practical psychology to acquire knowledge of the self or of the human spirit in general[2 3] , finds in this book by José Musangana a field of application for the acquisition of *Ubuntu* from nature, as an adjuvant that should enable us to ensure reciprocity towards this generosity of nature. In the context of this book, *Ubuntu* means :

"In the Xhosa language, Ubuntu means humanity. In Kinyarwanda, it means generosity. It is a concept that exists in almost all the Bantu languages of Africa. It encompasses the notion of community and interdependence between human beings. "[3]

In practice, Ubuntu has the following qualities:

---

[1] Nature - Google search
[2] introspective approach - Google search
[3] Ubuntu - Google Search

"Compassion, consideration, empathy, kindness, equality, human dignity and unity. "[4]

So we can say that we exist thanks to the existence of others and of Nature.

Remaining very practical in his approach, José Musangana proposes this approach, not as an open sesame to access an attitude valid for all, but as an instrument for community service, in a respectful and concerted process, with a view to a more conscious and consensual attitude.

Preserving nature today, so that human life can flourish, should be done by adhering to certain values and respecting ecosystems and the potential of nature. The approach developed in this book by José Musangana is consistent with this approach.

We enthusiastically invite potential readers to immerse themselves in this process of acquiring the Ubuntu of nature, with a view to conceiving nature as a vital territory with multiple vocations, requiring management that is both forward-looking and participatory.

Olivier-Jeff KALALA T.
Writer and literary critic
General Secretary of UECO

INTRODUCTION

There is no doubt that environmental disasters, climate change and the loss of biodiversity are caused by mankind, which has so far failed to create a climate of harmony between itself and Mother Nature.

Initially established by the Uncreated Creator as the steward of *Nature's* heritage[4] , man is often preoccupied with his own selfish interests and greed, without any concern for the fate of future generations. In this way, he illustrates himself as the executioner of *Nature,* using a capitalism with a barbaric face that is harmful to planetary equilibrium.

Multinational arms dealers are doing good business by plundering minerals around the world and, in this case, in Africa, more specifically in the eastern part of the DRC. In so doing, they are exploiting their economic interests by imposing their hegemonic motives.

This is the rationale behind bellicose conflicts around the world, where war remains the only way for many superpowers to do business or extend their hegemonic influence, with expansionist ideals, according to their interests. This is the case with the Israeli-Palestinian, Israeli-Iranian, Russo-Ukrainian, Rwandan-Congolese and Sino-Taiwanese conflicts, to name but

[4] Definitions: Nature - Larousse French Dictionary

a few.

The example of the DRC is too well-known to mention here. Nevertheless, it is no secret that it is one of the richest countries in the world. Unfortunately, its people are hardly benefiting from its natural resources, because of the hegemonic aggression of the multinationals that are exploiting its eastern neighbour, Rwanda. Observers see this as the black hand of the imperialists, who, by imposing a proxy war, are deliberately plundering the DRC's riches, which have been overwhelmed by nature's generosity. There are natural resources such as timber, oil and natural gas or methane, gold and diamonds, as well as minerals essential to the energy transition, such as cobalt, lithium and nickel. The Congo Basin forest in Central Africa is the second largest tropical forest after the Amazon rainforest. There are also peat bogs in the central basin of the Congo Basin, covering almost 145,500 km^2 , straddling Congo-Brazzaville and the DRC.

As well as cobalt, lithium and nickel, manganese is also necessary for the body to function properly.

Coltan (Colombo-Tantalite) is also recorded because of its particular ability to store and generate electrical energy; this mineral is used in mobile phones, laptops and other devices.

It is worth pointing out that the DRC holds the world's largest coltan deposit in marketable quantities, representing 60-80% of global reserves.

However, instead of making it the object of shared prosperity, under the win-win formula, between investors and the Congolese state, the imperialists, with their international firms, prefer to acquire it at the price of blood and in defiance of the principle of international law, namely *the intangibility of borders.* At the present time, an impressive number of people have been killed in the eastern part of the DRC .[5]

This situation is not without its detrimental effects on nature.

We are witnessing the excessive and massive slaughter of civilian populations, the disruption of ecosystems to such an extent that animals are forced to leave their natural environment, trees are crying because they are immobile, not knowing how to move, forests are devastated, rivers are polluted and bombing raids pollute nature...

All because of the identity crisis that man is undergoing.

In fact, fascinated by the material and thirsty *for possessions* and *power,* man has forgotten his "original being", and seems unaware that he is first and foremost nature in miniature, both physical and spiritual. In return, the only way to express his gratitude should be to protect the very nature from which he

[5] number of deaths in eastern DRC - Google search

sprang.

These questions are also worth asking:

Despite nature's generosity, why is it that mankind is always so belligerent towards nature?

What can be done to restore the natural order?

To do this, man should go through introspection to achieve reminiscence, in the Platonic sense, i.e. to access the memory of knowledge acquired in a previous life, when the soul, living in the suprasensible world of essences, contemplated ideas[6] , with a view to visualising and observing the rules of *Ubuntu,* written in bold letters on nature. Understood as humanity coupled with generosity in African languages, *Ubuntu* is the main attribute of the Uncreated-Creator, who deploys it through his vital energy over nature, which is the Space-Universe. In other words, the humanity of nature, *Ubuntu* of nature, refers to nature's capacity to act in a benevolent, compassionate and altruistic way towards the living beings that are part of it.

This manifests itself in phenomena such as the regeneration of ecosystems, the provision of resources essential to life, and the beauty and inspiration that nature offers to mankind.

Recognising and respecting nature's humanity is essential to preserving its balance and diversity.

[6] Platonic meaning of reminiscence - Google search

Our discussion will focus on 4 points:

1. Humanitarian Triad, to analyse the initial role of each of these three players: the Supreme Being, whom we identify as the uncreated-creator, Mother Nature and Man.

2. Nature's main feature: Ubuntu.

3. Causes of man's inhuman action against the generosity of nature.

4. Consequences of the dehumanisation of Nature.

CHAPTER ONE

HUMANITARIAN TRIAD

SECTION 1

CONCEPTUAL APPROACH

According to the Larousse French dictionary, a triad is a group of three people or three things closely associated. It can also refer to a group of three deities associated in a cult[7] . The triad we are talking about here is made up of heaven, earth and man. It is with this triad that we will try to support our argument.

Indeed, heaven refers to the Uncreated-Creator, while earth represents nature, of which man is the steward.

To understand man's ingratitude towards nature's generosity, we need to go all the way back to the Creator, to see how the latter created a vital space (nature), enabling man not only to manage it, but also and above all to live in it harmoniously and rationally. It is in this context that we will look at the role of each member of this triad: the Uncreated-Creator, nature and man. All linked by *Ubuntu,* a common identifier for each of them.

---

[7] Jeuge-Maynart, I., *Petit Larousse Illustré,* Larousse, Paris, p. 1170.

§1 HUMANITY

According to the LAROUSSE dictionary, the word *humanity* refers to all human beings, and is also considered to be a collective being or a moral entity.[8]

It is also a psychological disposition towards consideration, compassion, empathy and respect for fellow human beings. This is how Daniel Mulenda Lomena puts it: "Morality directs the subject towards others and invites him to behave with a view to promoting the lives of others".[9]

It is this second meaning that interests us. African cultures help us to understand the profound meaning of humanity, through the term *Ubuntu.*

§2 UBUNTU

The concept is defined along two distinct axes. Firstly, Ubuntu means being friendly, hospitable, generous, compassionate and attentive to others. In other words, putting your strength at the service of your neighbour, the weak, the poor, the sick, for no gain whatsoever. Treat them as you would like to be treated. The second is openness, magnanimity, sharing and recognition of the humanity of our neighbour. It is

---

[8] Jeuge-Maynart, I., *Petit Larousse Illustré,* Larousse, Paris, p.590

[9] Mulenda, D., *La gestion de l'intégration des entreprises par la préservation des écosystèmes naturels,* Paris, l'Harmattan, 2017, p. 100

inextricably intertwined with our own.

It is first and foremost an attitude steeped in ethics that encourages openness to others. Doesn't Corine Pelluchon say that the ethics of consideration implies a process of individualisation, linked to the relationship with others .[10]

In other words, the other is not simply the person who enables me to develop my potential, as in the relationship between master and disciple, but rather the person who constitutes my identity. This is achieved in a movement that is not self-referential, and is also distinct from the self-centredness that is characteristic of responsibility .[3]

§3. INCREE-CREATOR

The Uncreated-Creator, the Supreme Being, the first of beings, the initiator of life, all the more so as Nature and man were born from his fingers.

According to his plan of creation, the Supreme Being has foreseen that the existence of nature precedes that of man. This was done in order to create a living space for man, which he covers with Ubuntu. In short, the Supreme Being spreads his Humanity over Nature, endowing it with the capacity to act in a benevolent, compassionate and altruistic manner towards the

---

[10] Pelluchon C., *Éthique de la considération,* Paris, Seuil, 2018, p.84

living beings that are part of it, including Man.

§4 NATURE

It is creation par excellence, *already there,* we distinguish between space-Universe and space-Planet (earth), prepared in advance and whose capacity to accommodate is measured in immensity. Nature, as space-Universe, includes stars, planets, comets, asteroids, the sun and the moon. As for Nature as Space-Planet (earth), it includes flora, fauna, mountains, waters and many others. All this is the work of the Uncreated-Creator. In short, nature is a favourable environment that enables man to provide for his most basic needs, thus conditioning his survival. Above all, it is the place where *Ubuntu* manifests itself, which is its foundation.

§5 MAN

It is the creature endowed beforehand with responsible freedom, which gives it the rational capacity to judge, appreciate, discern and be guided by common sense. He is the *not-yet,* nature in miniature, all the more so because in him we find the 4 elements of nature (earth, water, air and fire). Although he was created after Nature, he becomes its manager established by the Uncreated-Creator. Since he is drawn from nature and benefits from its generosity, he carries within him the seeds of this beneficence, and is therefore intrinsically capable of *Ubuntu*, which is his original being.

It is important to make it clear that man does not humanise nature, otherwise this would be a usurpation of power. Man does not give meaning to nature; he is merely a microcosm that draws its source from the macrocosm that is Nature.

This is what justifies its third place in the hierarchy of the natural order (humanitarian triad), which includes: the Uncreated-Creator, nature and man.

This is why the existence of nature precedes that of man, because the Uncreated-Creator first wanted to create a space (nature) to shelter man, for his survival.

As a result, he is totally dependent on Nature, so that it is Nature that humanises man, looking after him in terms of health, science and technology, and stimulating his inspiration.

In return, man is invited to perpetuate the humanity of nature, of which he is largely the beneficiary, by preserving it. However, nature can continue to exist without man's contribution. In the words of George Perkins Marsh: "Away from man, nature cultivates itself; nature shapes its territory in such a way as to give it an almost unchanging permanence of form, outline and proportions, except when it is disturbed by geological convulsions. In such cases, which are relatively rare, she immediately sets about repairing the superficial damage she has sustained and restoring the land to its former appearance" .[11]

What's more, man's existence is indispensable in nature, as it is an opportunity for nature to express its characteristic generosity.

[11] Perkins G., *L'homme grand perturbateur de la nature,* Espace et Signes, Paris, 2023, pp 34-35

CHAPTER II

UBUNTU

Main characteristic of nature

Nature, the place par excellence for the manifestation of humanity, is at the service of man, in all his basic and complex needs. This is manifested in phenomena such as the regeneration of ecosystems, the provision of resources essential to life, and the beauty and inspiration that nature offers mankind.

Humanity as a whole depends on the health of natural ecosystems. In addition to their formidable carbon sequestration capacity, which is the primary natural solution to global warming.

These ecosystems have many useful functions: they provide habitats for plants and animals, shelter fish and other food sources, prevent erosion and protect against floods and other meteorological phenomena. When healthy, these ecosystems provide many essential benefits to mankind: they contribute, among other things, to economic resources, food, the preservation of water and land, the fight against drought, health and well-being... They are also a major source of the raw materials on which our economy is based.

The aim here is to describe the humanity or Ubuntu that

emerges from nature as Space-universe and Space-planet (earth) for the benefit of mankind.

SECTION 1

MANIFESTATION OF HUMANITY IN ALL ITS ASPECTS

§1. SPACE-UNIVERSE

When talking about nature, it would be incomplete to limit ourselves to planet Earth alone, because the works of the Uncreated-Creator are so grandiose that they include the billions of stars, the various planets, comets and asteroids. All these bodies are structured into galaxies, clusters and superclusters, dotted with such aesthetic and artistic beauty that they inspire and attract mankind. Such is the case with his journey to the moon. In other words, this work, which bears the personal stamp of the Great Architect of the Universe, is made available to man to arouse his curiosity. Man will therefore carry out research and sharpen his inspiration, in order to come up with new inventions in the artistic, scientific and technological fields. The fact that the Uncreated Creator is altruistic and generous towards man is equivalent to the humanity that freely places man in benevolent conditions for his fulfilment.

This is how we can fully enjoy the benefits of the Universe, whose matter is made up mainly of hydrogen and helium, whereas the Earth is made up mainly of oxygen, iron, silicon and magnesium. And living beings are made up of carbon, hydrogen, oxygen and nitrogen.

It's worth pointing out that, according to scientists, of the 118 chemical elements that make up matter in the Universe, 94 exist on Earth.

§2. SPACE-PLANE

Before bringing out the humanity that Nature contains for the benefit of mankind, it is imperative to define certain related concepts.

A. BIODIVERSITY

Biodiversity refers to the variety of life forms on Earth. It is assessed by considering the diversity of ecosystems, species and genes in space and time, as well as the interactions within and between these levels of organisation. Although biodiversity is as old as life on Earth, the concept only emerged in the 1980s.

B. ECOSYSTEM

An ecosystem is an assembly of animals, plants and micro-organisms interacting with each other and with their environment. These interactions take place in systems that are more or less natural: forest, lake, agricultural, urban, etc.

1. **Different types of ecosystem**

- Terrestrial ecosystems include forests, deserts, grasslands, mountains and polar zones.

- Marine ecosystems (oceans, seas, coral reefs, estuaries)

- Freshwater ecosystems (lakes, rivers, wetlands).

- Wetland ecosystems, such as mangroves

2. **Difference between** :

   - A natural disaster,
   - A natural phenomenon
   - An environmental disaster

A natural disaster is a violent natural event with destructive effects on the earth's surface, on humans and on other living beings. It is in fact the manifestation of a natural hazard. E.g. volcano, earthquake, tsunami.

A natural phenomenon is an observed natural fact that can be studied scientifically and used as a subject for experiments: meteors are natural phenomena. In short, a surprising manifestation of nature, of climatic origin.

An environmental disaster refers to a disaster caused by man himself. E.g. global warming.

SECTION 2

THE HUMAN SIDE OF NATURE IN THE LIFE OF A MAN

Nature has never ceased to show its generosity towards mankind. In terms of food, health and even microbiology.

§1. FOOD

Nature, the source of balanced food, offers the choice of a diversified diet. In this way, the body benefits from a varied source of nutrients that are essential to its functioning.

By eating a wide variety of fruit and vegetables, we can avoid vitamin and mineral deficiencies that weaken the body. Diversity is essential to our health! By the same token, a diverse range of crops slows down the spread of disease (viruses, bacteria, fungi, etc.) and limits the damage to crops caused by certain animals that have become veritable pests (insects, mammals, birds, etc.).

§2. HEALTH

- Biodiversity, our pharmacy germoir, natural skincare.
- Animals treat themselves with natural remedies. They could be the source of new medicines for humans.

Grandmother's medicine, traditional medicine or pharmacopoeia. People have always drawn their remedies from nature. Even today, a large proportion of the world's population still relies on these often less costly practices, whose value is recognised by the World Health Organisation (WHO). But some plants are under threat, and traditional medicines are disappearing before they have even been the subject of in-depth ethno-pharmacological studies.

- Medicines from nature

More than half of all medicines come from nature. The molecules that have a therapeutic action are either used directly in medicines or, more often than not, isolated and then synthesised in the laboratory. Today, the manufacture of medicines is less dependent on natural resources, but these are still very precious to our health.

§3. AT MICROBIOTA LEVEL

Microbiota, a group of microorganisms: bacteria, viruses, parasites and non-pathogenic fungi, known as commensals, which live in a specific environment.

There are different microbiota in the body: in the skin, mouth, armpits, vagina, lungs, etc.

Our bodies are home to a wide variety of genes, as well as bacteria, viruses and fungi. Frequent contact with nature exposes us to new microbes that strengthen our macrobiota and immune system. This could help prevent the onset of asthma, allergies or certain inflammatory or autoimmune diseases.

Despite this benevolence of nature, man excels in ingratitude. But what drives him to act in this way, one might ask.

CHAPTER III

CAUSES OF DEHUMANISATION OF NATURE

This chapter looks at the causes of man's ingratitude towards the delights of Nature.

This approach involves analysing man in two dimensions: original being and temporal being.

SECTION 1

ANALYSIS OF THE NATURE OF MAN

§1. MAN, THE ORIGINAL BEING

As the original being, man is predisposed to communion with nature. He is created with every capacity to be benevolent towards nature and his fellow man. He is made of Ubuntu. This is how the Uncreated-Creator established him as the steward of *nature's* heritage, so that he could be the promoter of humanity and perpetuate life around him. In addition to his original state, he enjoys a responsible freedom, equipped with common sense and the spirit of discernment, knowing how to distinguish between good and evil. He is in a manifest dimension of humanity. This responsible freedom is justified by the fact that the Supreme Being did not want a puppet or a sheep of Panurge. But rather a man who is aware of his responsibilities, compassionate and benevolent.

At this level, man is the prior beneficiary of nature's benevolence. Nature expects us to continually respond in terms of protection and preservation.

At this stage, the humanitarian triad takes the following form: the Increate-Creator begins by providing a space in which man can live. This is nature, with everything man needs for his survival, both food and sanitation - in other words, the vital resources that are essential for his survival! For his part, man signs a contract of guardianship with the Uncreated Creator, establishing him as the guardian of the temple. This is where he learns to play the balancing act. He is so deeply rooted in his dimension of original being that he lives in the precomprehension of a humanity that wants to be manifest in him, so that benevolence, generosity and altruism go hand in hand.

However, in his original state, man is fully human. But he learns the rudiments needed to carry out his mission. In other words, he must put his humanisation into practice once on earth. To do this, he must first be confronted with the trials of time.

§.2. MAN, THE TEMPORAL BEING

Thrown back in time, man has forgotten his original role as manager of *Nature's* heritage and promoter of humanity. He is now suffering from an identity crisis. In reality, man suffers from the predominance of his temporal being, which forces him to terminate the contract of guardianship that binds him to Nature. He falls into a latent dimension of humanity. Compared to Nature, the *already-there,* therefore, fully human by virtue of its generosity, man, the *not-yet-there,* since, fascinated by the evolution of society which is very materialistic, forgetting that he was created to guard the temple, to manage all that is at his disposal, in a rational, equitable and balanced way and above all for the benefit of all. He becomes disconnected from his original being.

He is struck by the amnesia of the materialistic excess that characterises today's society. It becomes thirsty for possession and power. As a result, the spirit of competition and hegemony takes hold of him. Now enjoying absolute freedom, leading to libertinism, man's behaviour is liberticidal towards Nature and his fellow man.

So any means are good enough to satisfy its selfish interests and voracious appetites. Its initially responsible freedom becomes absolute, massacring innocent civilian populations as a result, in the name of capitalism with a barbaric face, unfortunately unable to achieve prosperity for all (Win-Win). The same applies to international companies which, by imposing wars in Africa and throughout the world, through certain NGOs which transport heavy weapons and munitions, kill without faith or law. The so-called high-risk zones, under the guise of humanitarian corridors, are a strategy used to plunder the natural resources of the cradle of humanity and of other countries, and to sell their weapons to the highest bidders.

By acting in this way, man breaks the harmony that should reign between himself and nature, dehumanising himself by killing his fellow man, who is also part of nature. In turn, he attacks nature itself, committing genocide by seeking to monopolise natural resources in a bellicose and illicit manner. It should be pointed out that this attitude has devastating effects on the environment. Massacres perpetrated for economic or survival purposes are tantamount to crimes against humanity, but also crimes against Nature.

§.3. MAN'S ACTIVITIES AGAINST NATURE

To ensure our survival, we engage in a host of activities. But unfortunately, his selfishness distracts him from his responsibility to future generations. His use of unorthodox means to satisfy his baser instincts unconsciously leads him to create a collective misfortune, such as global warming, wrongly considered a natural disaster.

A. TYPES OF ACTIVITY

- Agricultural activities (with the use of pesticides), the exploitation of the land and livestock farming (based on GMOs). These practices are slowly contributing to the extermination of mankind.

In the same vein, Rachel Carson warned: "Apart from the risk of exterminating humanity through atomic warfare, the crucial problem of our time is the contamination of our environment by incredibly harmful substances. These products accumulate in the tissues of plants and animals and even penetrate reproductive cells, where they alter the elements that determine the future through heredity".[12]

- Industrial production at the root of pollution.

Industrial pollution can take different forms. Firstly, there is air pollution from the smoke discharged by factories. Then there is soil and water pollution from wastewater and industrial waste. Finally, there is noise pollution caused by industrial activities.

As a result, sewage and waste disrupt the nitrogen cycle in the soil and cause salt to build up in the soil, making it sterile and leaving no vegetation to grow.

Just as rivers are polluted by mining activities. Nature's humanity must be taken into account before, during and after

[12] Carson, R., *Silent Spring,* Wild Project, 2009, p.50

any human undertaking.

Article 203 of the DRC Mining Code, which deals with the protection of the environment during exploration, states that "before commencing exploration work for minerals or quarry products, the holder of an exploration permit or an authorisation to explore for quarry products must draw up and obtain approval for a mitigation and rehabilitation plan for the proposed activity".[13]

And during operation, article 204 stipulates that "any applicant for an operating permit, a discharge permit, a small-scale mining permit or a quarrying permit is required to submit an environmental and social impact study, together with an environmental project management plan, and to obtain approval for its ESIA and PGEP and implement the PGEP".[14]

As for damage caused to people and the environment by contamination, the holder is liable under article 285ter: "in the event of direct or indirect contamination as a result of mining activities, having an impact on human health and/or leading to environmental degradation and resulting in particular in the pollution of water, soil or the atmosphere and causing damage to people, fauna and flora".[15]

---

[13] *Code Minier révisé et annoté de la République Démocratique du Congo,* Bruylant, Brussels, 2020, p.242

[14] *Code Minier révisé et annoté de la République Démocratique du Congo,* Bruylant, Brussels, 2020, p.243

[15] Ibidem, p.332

- Deforestation

By cutting down forests, mankind is working to extinguish the species that live in them. We are destroying ecosystems that produce drinking water, and we are losing the trees that store CO_2. This further alters the climate. George Perkins Marsh adds: "the destruction of forests leading to soil erosion, the draining of lakes and marshes, agriculture and industry have tended to produce major changes in the hygrometric, thermometric, electrical and chemical conditions of the atmosphere, although we are not yet able to measure precisely the share of each of these different elements of disturbance".[16]

Despite the beneficial effects of ecosystems on human life, we continue to abuse them, deforesting and degrading forests, and resorting to intensive production methods that impact on the climate and have a knock-on effect on local populations and biodiversity. In so doing, mankind is not only causing the scarcity of certain natural resources, but also disrupting the economy of certain businesses. This is a major drag on the development process for many countries, preventing them from emerging. Added to this are the horrors of war. Let's talk about that.

- War, dehumanising Nature

[16] Marsh, G. P., *L'homme grand perturbateur de la nature,* Espace et Signes, Paris, 2023, p.10

Human extermination, for whatever reason, joins the chain of dehumanisation of Nature, because the exterminator and the exterminated are part of Nature. This is why killing a person is a crime against the humanity of Nature. This is more encompassing than the crime against humanity, especially as man is a microcosm among many others forming part of the macrocosm (Nature). In other words, war affects not only man, but above all the humanity of Nature as a whole. As a result, the blood that flows disrupts ecosystems and mourns the loss of Nature's members, due to man's barbarity, which has reached its peak.

It is to be deplored that, for the Powers of this world, the use of war is becoming a means of plundering weak, albeit potentially rich, countries. This new conception of international relations had never ceased to be decried, throughout the world and at international assemblies.

B. THE CASE OF THE DRC

At the Three Basins Summit, alongside his counterparts Denis Sassou Nguesso of Congo and Lula Da Silva of Brazil, the President of the DRC, Félix Antoine Tshisekedi, gave a speech that was greatly appreciated by the audience.

On 29/10/2023 in Brazzaville, President Tshisekedi denounced Rwanda's armed activism, instrumentalised by occult powers pulling the strings, in destroying Congolese biodiversity in Virunga Park. In a firm tone, he clearly accused the Kigali regime of plundering Congolese mineral wealth.

Addressing the audience, which included heads of state and official delegations, representatives of governments, international institutions, donors, funding bodies and experts from around the world, Félix Tshisekedi recalled that one of the world's most precious biodiversity reserves, Virunga National Park, had been damaged by the continuing conflict in the east of the DRC. This war is seen as a factor in the dehumanisation of nature in the eastern part of the DRC.

At the root of it all was an economic conflict between the DRC and Rwanda. Conflicts of this kind have occurred before. However, all it took was an iron will to bring about a peaceful resolution. The famous European Coal and Steel Community was established in April 1951 to promote economic cooperation and prevent conflict between European nations after the Second World War. For these two countries in the Great Lakes region, it would be useful for Rwanda to lay down its arms in favour of peace, which is highly conducive to the preservation of ecosystems. Under these conditions, it could ask the DRC to be one of the commission agents, whose percentage is generally deducted at source during mining transactions. This would require mediation by an institution more serious than the international community, which has repeatedly failed to resolve conflicts peacefully. It is often at the mercy of the Great Powers, which use it to crush other countries.

This attitude on the part of the international community, which is limited to verbal condemnation, is becoming a cause for concern. It is deplored in the Israeli-Iranian conflict, which risks opening the door to the third world war because Israel, which has the support of the USA, is not unaware of the support of the Arab countries for Iran, which already enjoys Russian support. And Russia has military agreements with China and North Korea. This will be an opportunity for China to attack Taiwan and for North Korea to settle its score with South Korea. Imagine the rest...

That being the case, it is essential to emphasise man's indifference to the ecological risks of which he is the author. Instead of making it a priority, we opt for the massive destruction of our environment and our fellow human beings, through wars that expose our barbarity. Hence Legor Gran's question: "Will we continue to multiply or will we opt for nuclear war?[17]

All these inhuman actions by man call for sanctions by nature.

[17] Gran L., *L'écologie en bas de chez moi,* P.O.L, 2011, pp 120-121

CHAPTER IV

DEHUMANISATION

OF NATURE

The consequences of man's dehumanisation of Nature can be seen as a self-inflicted punishment.

They manifest themselves in a number of ways.

SECTION 1.

REPERCUSSIONS OF DEHUMANISATION

- Environmental disasters,
- Climate change,
- Loss of biodiversity,
- The rise in average global temperature,
- Climate change, which is responsible for extreme weather phenomena such as heat waves and droughts,
- Water and food shortages,
- Bushfire,
- Land subsidence due to the haphazard mining of minerals, etc.

What measures are envisaged to re-establish the natural order?

SECTION 2.

RESTORING BALANCE

NATURAL

This chapter is structured on four levels: firstly, it proposes a number of possible solutions, including meditation as the royal road, through which man can take an in-depth look at himself in order to rediscover his original being. Then there will be a discussion of ecological exercises to encourage communion with Nature. In addition, there will be the need for a school curriculum based on the protection and preservation of Nature. Finally, the community will have to impose a penalty on any recalcitrant person who, whether consciously or not, repeats the crime against Nature's humanity, offering the image of a being without faith or law.

§1 MEDITATION

Communion with nature

In order to provide for their needs, man seizes the opportunity offered by the Uncreated-Creator and exploits the delights of nature. But, irrationally, man often goes so far as to damage the natural balance.

Is it possible to prevent such an attitude?

Made up of two dimensions, the original being and the temporal being, it is plausible for man to withdraw into himself, because the solution to the problem lies in immanence and not in transcendence. Hence the need for profound meditation, first creating a void, with a view to ridding ourselves of everything that is linked to temporal accessories and selfish interests, to the detriment of the community.

In the same vein, Corine Pelluchon adds: "Most of the time, we don't feel that we are in community with other beings, or that we belong to a vast universe, because we have a partial perception of ourselves, we think and live by relating our sensations and emotions to our ego, and so we imagine that everything that exists must serve our selfish interests.[18] This is why deep meditation is needed to enable us to encounter ourselves. It is an impulse towards his original and total dimension. This meditation will lead to "reminiscence"[19] . This term is central to Plato's philosophy. It refers to the memory of a previous state, when the soul possessed a direct view of ideas.

Reminiscence is therefore the foundation of man's power of knowledge. It is at this level that man will finally be able to contemplate his original being and re-enter the contract of guardianship signed with the Creator, with a view to sharpening his altruism, his generosity and his capacity as a rational manager of nature. In doing so, he will readily think of his community. He will therefore avoid seeing nature as his victim,

---

[18] Pelluchon C., *Éthique de la considération,* Seuil, Paris, 2018, p.88

[19] Plato, *The Menon,* the Phaedo and Phaedrus

but rather as an alter ego, a partner with whom to exchange benefits and ensure common protection. This is man's original mission, Ubuntu.

Re-learning all this knowledge, and putting it into responsible practice, will harmonise the relationship between man and nature. Aldo Leopold had already understood this when he said: "Protecting nature leads to a state of harmony between man and the earth".[20]

It is this approach that will help them to rediscover their status as human beings, which, according to Corine Pelluchon, means "being part of a community of humans characterised by plurality".[21] This is how man will be fulfilled, moving from *Not-yet* to *Already-there,* while at the same time reappropriating human nature, Ubuntu. He will no longer be able to use his absolute freedom over nature, but it will be a responsible freedom regained. Clearly, changing ingrained habits is not something that can be done overnight, but a series of exercises will help you to achieve it.

§2. ECOLOGICAL EXERCISES

(Eco-gestures)

It's always useful, beneficial and favourable for the body and

[20] Léopold, A., *L'éthique de la terre,* Payot, Paris, 2019, p.23
[21] Pelluchon, C., *Les lumières à l'âge du vivant,* Seuil, Paris, 2021, p.116

mind to be in contact with nature. Walking in the trees, in the garden, or contemplating, admiring the scenery and letting yourself be amazed by nature... these are all exercises that are good for human mental health. The heart, muscles and senses also benefit. As a result, functional disorders such as depression, stress and anxiety are significantly reduced. As you can see, nature regulates your mood and helps you concentrate when you're learning.

Environmental exercise, which is essential at all ages, makes it easier to meet other people and boosts the immune system. There's nothing like a breath of fresh air, especially when you lead a sedentary lifestyle!

There's nothing more soothing than sowing or planting. This phenomenon can be explained by the fact that the soil contains a bacterium with beneficial effects. When you touch it or breathe it in, the production of serotonin (a chemical messenger that some call the happiness hormone) is activated in the brain. Gardening stimulates the senses. When we come into contact with flowers or aromatic plants, we remember the smells or the dishes we loved so much. That's why therapeutic gardens have been created for the elderly, the disabled and people suffering from neurodegenerative disorders such as Alzheimer's disease, Parkinson's disease, Lewy body disease and other dementias.

Playing and learning in the garden, touching, smelling and

tasting what you've planted, is an opportunity to commune with Nature.

§3. PROGRAMME INNOVATION SCHOOL

(On nature protection)

Being close to nature to get to know it and ourselves better is a programme that needs to be incorporated into education systems throughout the world, because the earlier children learn to treat nature humanely, the less damage there will be.

That's why it's so wonderful to see children climbing trees, collecting dead leaves and observing the little creatures that flourish...

Teachers who regularly take their pupils out into nature find that they become more agile and attentive than before, and that they memorise more easily. We agree with George Perkins Marsh, who stresses the importance of education in understanding and appreciating our environment and in promoting sustainable practices to preserve it[22] . Within the family too, nature experiences, such as those organised by environmental organisations, are highly beneficial for both children and adults. However, once you have gone through this

[22] Perkins G., *L'homme grand perturbateur de la nature,* Espaces et Signes, Paris, 2023, pp 11-12

process of reconciliation with nature, which, incidentally, requires the will of man. There is no room for error. If they do so again, they will suffer the appropriate penalties.

§.4. PROPOSAL

CRIMINAL PENALTIES

The international community should unanimously stand up and severely punish acts that dehumanise nature.

This is the case with blood minerals and wars around the world. Nature wants to live in peace, in a climate conducive to the balance of ecosystems.

This is why a world court specialising in the humanity of Nature should be set up. Its operation will sanction the errors of certain United Nations systems, whose method favours the impunity of major Powers that are also major polluters.

The JMSHN (World Jurisdiction on the Humanity of Nature) will be able to judge recalcitrant perpetrators of crimes against the humanity of nature, and should not be influenced or intimidated by certain powerful countries. It should also deal with cases of massacres of civilian populations, motivated by the plundering of natural resources.

The DRC's reaction is to be encouraged, as it should be successful.

As reported in the newspaper Le Monde, the DRC has

accused Apple of using *illegally* mined minerals in its products, allegedly *from Congolese mines* where *many human rights are violated,* according to documents consulted by Agence France-Presse (AFP).

According to the lawyers instructed by the DRC, these minerals are then *transported out of the DRC, in particular to Rwanda, where they are laundered.*

In addition, it appears from the DRC's file, submitted to the ad hoc bodies, *that Apple uses strategic minerals purchased in Rwanda in its products,* according to the lawyers instructed to draw up a formal notice, a summons before legal proceedings are initiated.

"Rwanda is a central player in the illegal exploitation of minerals, and in particular the exploitation of tin and tantalum from the DRC", say the lawyers. "*After being extracted illegally, these minerals are smuggled into Rwanda, where they are integrated into global supply chains",* the formal notice states. "*These contentious minerals are said to come largely from Congolese mines, where many human rights are violated",* the lawyers continue.

This formal notice has been sent to Apple's two subsidiaries in France by French lawyers William Bourdon and Vincent Bren Garth. A letter has also been sent to the American parent company of the major technology group, which markets the iPhone and Mac computers, among other products.

CONCLUSION

Our approach has been to look deep inside man to understand why he acts so inhumanely towards nature.

To this end, we have presented the humanitarian triad in order to grasp the roles of each actor, through the attitude to be observed, with regard to the management of Nature. It emerges that Nature is covered by Ubuntu, the main attribute of the Uncreated-Creator, who establishes man as the manager of Nature, in its original state, on the basis of a contract of guardianship. But once on earth, man is struck with amnesia, due to the radiation of materialism that characterises society. As a result, we behave as destroyers of nature, killing our fellow creatures because of the selfish interests and voracious appetites of certain international corporations. Hence the need for man to make an inner journey, through deep meditation and reminiscence, to help him rediscover his original being, his humanity, without which he risks disappearing because of the wrath of nature. This is one way of combating the dehumanisation of nature.

Man's failure to respect ecological life has obituary consequences for the whole of humanity.

BIBLIOGRAPHY

Carson, R., *Silent Spring,* wild project, 2009

Code, *Code minier, révisé et annoté de la RD Congo, Bruyant, Brussels, 2020, 385p.*

Gran, L., *L'écologie en bas de chez moi,* P.O.L, 2011

Léopold, A., *L'éthique de la terre,* Payot, Paris, 2019

Mulenda, D., *La gestion de l'intégration des entreprises par la préservation des écosystèmes naturels,* Paris, l'Harmattan, 2017

Pelluchon, C., *Éthique de la considération,* Paris, Seuil, 2018

Les lumières à l'âge du vivant, Seuil, Paris, 2021,

Perkins, G., *L'homme grand perturbateur de la nature,* Espace et Signes, Paris, 2023

WEBOGRAPHY

1) Definitions: Nature - Larousse French Dictionary
2) number of deaths in eastern DRC - Google search
3) Platonic meaning of reminiscence - Google search

TABLE OF CONTENTS

MIX
Papier aus verantwortungsvollen Quellen
Paper from responsible sources
FSC® C105338

Printed by Books on Demand GmbH, Norderstedt / Germany